Management Division 2019

Held at the 2019 AIChE Annual Meeting

Orlando, Florida, USA
10-15 November 2019

ISBN: 978-1-7138-0529-8

Printed from e-media with permission by:

Curran Associates, Inc.
57 Morehouse Lane
Red Hook, NY 12571

Some format issues inherent in the e-media version may also appear in this print version.

Printed with permission by Curran Associates, Inc. (2020)

For permission requests, please contact AIChE
at the address below.

AIChE
120 Wall Street, FL 23
New York, NY 10005-4020

Phone: (800) 242-4363
Fax: (203) 775-5177

www.aiche.org

Additional copies of this publication are available from:

Curran Associates, Inc.
57 Morehouse Lane
Red Hook, NY 12571 USA
Phone: 845-758-0400
Fax: 845-758-2633
Email: curran@proceedings.com
Web: www.proceedings.com

TABLE OF CONTENTS

580336 A Chemical Engineer's Overview of Artificial Intelligence and Machine Learning

This hands-on workshop will start with a high level overview of Artificial Intelligence and Machine Learning, citing case studies of their development, and guide participants through identifying where these concepts can be applied in their chemical engineering practice.

How Virtual Advisors Create Practical Employee Efficiency Boosts in the Chemical Engineering Workplace

Alec Walker, DelfinSia, Inc, Houston, TX, USA.

Abstract

In an effort to explore the ability of text analytics software to augment human performance within the chemical engineering domain, a virtual advisor designed for the domain was used by chemical engineering practitioners in a controlled experiment. The experiment focused on the common industry problem of inefficient and incomplete extraction of relevant knowledge and data from within the organization's digital unstructured textual (DUT) files in order to resolve issues. The experiment focused on hydrocarbon refinery content within the broader chemical engineering domain. The virtual advisor used in the experiment employs machine learning and natural language processing to interpret the intent of human queries and instantly return knowledge and data from across multiple DUT files that is relevant to the queries. The advisor has shown capacity in commercial applications within the industry, but the key details of those results remain proprietary to the corporate clients.

While this study cannot name the individual participants in the experiment, the corporations they represent, or their proprietary issues explored using the virtual advisor during the experiment, the experimental results comparing the performance of the individuals with and without the virtual advisor are shared in detail. The results suggest that the virtual advisor increases the speed and accuracy of issue resolution in the chemical engineering domain over typical incumbent tools in a base case control scenario by significant margins. This in turn suggests that chemical engineering is an appropriate domain for the productive application of virtual advisors based on "artificial intelligence" methods including machine learning and natural language processing. The implications of the results are explored, and additional larger and more open studies in this area are recommended.

Introduction

The process-driven nature of the chemical engineering domain means that knowledge and data potentially useful in resolving issues already resides in the organization's internal documentation. When issues arise that demand resolutions, practitioners first use incumbent methods to extract relevant information from these documents before then applying their skills putting it to use. One incumbent method is to store files in hard copy form in a physical library, creating logistical challenges and risks. Leading organizations have digitized their files on shared folders. They often use a combination of inter-file methods, intra-file methods, and human reading to find data and knowledge buried in the organization:

1) The most common inter-file method is keyword-based enterprise search. This type of search tool relies on keywords to find a subset of the total files in the organization that contain the keywords entered by the practitioner into a search field. The practitioner then browses through the files in the search results, opening whichever seem relevant individually.
2) The most common intra-file method is keyword-based textual search. This functionality tends to be built into the software handing the individual file. When a potentially useful document found from inter-file search is opened, the practitioner typically has the option to use "Ctrl + F" to type in the key words again and browse through each of their occurrences within the document one at a time.
3) The final step in the incumbent process is simple human reading. Following the intra-file keyword search, if the document seems relevant, the practitioner reads it carefully looking

for useful knowledge and data. If the document does not seem relevant, the practitioner closes it and continues to browse the search results from the enterprise search.

Processes like these account for 80% of knowledge worker time in the energy industry (1), and that percentage is likely at least as high across the entire chemical engineering domain. Even though the relevant knowledge and data are contained in the organization's files, and despite the fact that the job of the knowledge workers is to put this information to use, locating and extracting the information still accounts for the vast majority of professional time (2).

Recent advances in the field of natural language processing now allow computers to perform much more advanced search than that based simply on keywords. The cutting edge in natural language processing, however, still requires domain-specific applications of the technology in order to be effective. As a result, domains are each beginning to separately experiment with natural language processing to explore its potential (3). With the goal of exploring the potential of natural language processing in the chemical engineering domain, a virtual advisor based on natural language processing was evaluated against an expedited incumbent method as described above. The virtual advisor takes in domain-specific questions rather than key words. The goals of the advisor are to reach faster and more accurate resolutions to the common issues requiring reference to digital unstructured textual (DUT) files. The virtual advisor attempts to accomplish these goals through a single search across and within multiple DUT files and by comparing each option within the digital text to the intent of the question asked.

Another recent study with a team of engineering consultants at a major multinational chemical company showed that this virtual advisor was able to successfully mine DUT files to extract information for practitioners more efficiency and effectively than their incumbent solutions. According to one engineer involved, "[The tool] helps me solve the problem of being able to search for information within reports which I already know contain that information but can't remember which report or where in the report." This study follows that industry trial for the purpose of creating detailed and accurate data open to the community pertaining to the effectiveness of the virtual advisor as a proxy for the potential of text analytics in the chemical engineering domain generally.

Key Terms

"Artificial Intelligence" – A popular term describing techniques in data science and computer science.

Machine Learning – A method in the field of computer science involving algorithms that iteratively adjust models to fit data.

Natural Language – What scientists call the form of language humans use to interact with each other as opposed to what humans use in traditional interactions with machines. Natural language does not imply English language.

Natural Language Processing – The science of using computers to process natural language. This could involve using algorithms to process audio files to convert speech into text, or it could involve using algorithms to process video files to read the movement of lips. This paper focuses on text analytics.

Text Analytics – The branch of natural language processing dealing with digital text written in natural language.

Digital Unstructured Textual (DUT) Files – These are files on computers that humans use to save and reference information. They are almost always written by a human, and they are composed of textual information. Typically, these files are saved on a hard drive or a shared folder. They might be reports, emails, memos, compliance documents, manuals, job descriptions, etc. They might end in such extensions as .ppt, .doc, .txt, .pdf, etc. This paper, when opened on a computer, is a good example of a DUT file. Note that some DUT files are mere images of typed text as opposed to actual typed text, adding one more layer of complexity to the process of employing text analytics. DUT files make up 90% of the total information across all organizations.

Methodology

The methodology involved a focus group experiment. Research and design chemical engineers from the hydrocarbon refining industry proposed resolutions to common workplace issues through reference to a set of DUT files in two scenarios: a base case using expedited incumbent reference tools and a text analytics case using a custom-built virtual advisor. The speed with which results were obtained was measured across both scenarios quantitatively (rounded to the nearest half minute), and the effectiveness of the solutions obtained was determined qualitatively from surveying the chemical engineers following the experiment.

The corpus of DUT files used in the experiment came from an open online forum for hydrocarbon refinery operations engineers started in 2009. The forum contains more than 6,000 DUT files in the form of typed discussion threads. In the base case, expedited incumbent reference tools for the experiment included a standard keyword search option and a browsable window showing all content from multiple DUT files. The standard keyword search option enables entering a keyword to have the system highlight that key word the entire body of DUT content. This effectively already cuts out the inter-file search step, giving the base case scenario an advantage over the typical scenario within a chemical engineering organization. In the text analytics case, the virtual advisor served as the only reference tool. The virtual advisor uses text analytics to interpret the intent of questions typed by the chemical engineer. The virtual advisor then processes the entire body of content within the DUT files and returns relevant passages to the chemical engineer all at once in the same window.

Seven chemical engineers with a combined total of 29 years of professional experience were involved in the focus group experiment. Four had backgrounds primarily in process improvement research, and three had backgrounds primarily in reactor design. Each were asked individually to consider their professional experience and come up with four issues they have faced which required reference to DUT files in order to be resolved. These issues were summarized generally, without specifics of the companies, reactors, or other details. For example, "The amine unit began producing problematic foam", and "The ratio of the height to the number of trays needed to be optimized". Despite anonymization of the identities of the chemical engineers involved with the study, the individual issues were deemed proprietary and are unable to be shared in this paper.

The chemical engineers were then sorted into groups based on their stated lack of professional experience with each of the issues. Seventeen of the issues had not been experienced by at least three of the seven chemical engineers. For a randomly selected nine of these seventeen issues, the members of the corresponding group lacking related experience underwent a trial using the base case scenario to craft a resolution for the issue. For the other eight of these seventeen issues, the members of the corresponding group lacking related experience underwent a trial using the text analytics scenario to come up with a resolution for the issue. We called these people "Proposers". For each trial, a time limit of ten minutes was imposed.

The proposed resolutions to the issues were then scored on their relevance and accuracy by the chemical engineers who mentioned having prior experience with these issues. We called these people "Evaluators". Comparisons between the scores obtained from each of the two scenarios were then made and explored. Concerning relevance of a proposal, evaluators were asked, "Is the proposal related to the issue?". Concerning accuracy of a proposal, evaluators were asked, "Could the proposal potentially be useful in resolving the issue?". The evaluators used the scoring system in Table 1 to respond to these questions in the case of each issue + proposal pair.

Table 1. Scoring System for Proposed Resolutions from Experiment Scenarios

Score	0	1	2	3
Relevance	No Relevancy	Little Relevancy	Mostly Relevant	Entirely Relevant
Accuracy	No Accuracy	Little Accuracy	Mostly Accurate	Entirely Accurate

Notes:

- Relevancy and accuracy scores of trials in which no proposals were made were unassigned (marked as "NA") rather than scored zero.
- Any proposal with a score of at least one was considered "relevant" in the resulting tables.
- Accuracy scores of proposals with a relevancy score of zero were unassigned (marked at "NA") rather than scored zero.
- Average percentages for relevance and accuracy were based on the range of options within Table 1. For example, an average accuracy score of zero was deemed 0% accurate, an average accuracy score of one was deemed 33% accurate, an average accuracy score of two was deemed 66% accurate, and an average accuracy score of three was deemed 100% accurate.

Results and Discussion

The results for the base case scenario show a relatively lower proposal rate, average relevancy, and average accuracy, as well as a relatively high average amount of time to research and craft a proposal.

Table 2. Scoring Summary for Base Case Scenario

Issue #	Proposers	Proposals	Avg. Time (Min)	Evaluators	Avg. Relevancy	Avg. % Relevancy	Relevant Proposals	Avg. Accuracy	Avg. % Accuracy
1	3	3	9.5	4	0.33	11%	1	2	67%
2	4	3	10	3	1.33	44%	3	1	33%
3	4	0	10	3	NA	NA	0	NA	NA
4	4	1	9.5	3	0	0%	0	NA	NA
5	5	5	7.5	2	2.2	73%	5	1.6	53%
6	3	3	10	4	1	33%	1	1	33%
7	3	3	2	4	3	100%	3	2.33	78%
8	3	0	10	4	NA	NA	0	NA	NA
9	3	3	8	4	1	33%	2	1	33%

The results for the text analytics scenario show a relatively high proposal rate, average relevancy, and average accuracy, as well as a relatively low average amount of time to research and craft a proposal.

Table 3. Scoring Summary for Text Analytics Scenario

Issue #	Proposers	Proposals	Avg. Time (Min)	Evaluators	Avg. Relevancy	Avg. % Relevancy	Relevant Proposals	Avg. Accuracy	Avg. % Accuracy
10	3	3	4	4	2	67%	3	1.67	56%
11	4	4	2	3	2.75	92%	4	2.5	83%
12	4	4	3	3	2	67%	4	2.75	92%
13	3	2	7	4	3	100%	2	1.5	50%
14	4	4	4	3	3	100%	4	1.25	42%
15	3	3	4	4	2.33	78%	3	1.33	44%
16	4	4	2	3	3	100%	4	2.5	83%
17	3	1	10	4	0	0%	0	NA	NA

In the base case scenario, there were 32 total opportunities for proposals, but there were only 21 proposals (66% of total opportunities) made by the end of the ten minutes in each trial. In the text analytics scenario, there were 28 total opportunities for proposals, with 25 proposals (89% of total opportunities) made by the end of the ten minutes in each trial. There are two possibilities for each case of these discrepancies: either the time limit was too short for the chemical engineer to find relevant information to formulate a proposal, or the DUT files lacked relevant information. While lack of relevant information was reportedly a factor, twelve proposers reported running out of time and having to submit no proposal or half-formulated proposals in the base case scenario, whereas only two proposers reported running out of time in the text analytics scenario. The text analytics scenario also shows a smaller average time requirement for proposal crafting (4.5 minutes across all proposal opportunities for the scenario) than the base case scenario (8.5 minutes across all proposal opportunities for the scenario). These results strongly suggest that the virtual advisor was able to drastically speed up the process of crafting resolutions to common issues. One participant explained, "I just typed the issue into the virtual advisor interface and then instantly found all the relevant information from the files. The advisor put all the answers from the files on the screen for me so I could just read them and combine them into my answer each time…The keyword search makes me scroll through each result one at a time, and most are not actually relevant to the issue at hand. It's much slower."

These results are important to consider in the context of a typical working environment. The base case scenario treated the body of DUT files as the equivalent of a single DUT file, cutting out the inter-file search and instead allowing the chemical engineer to jump to instances of keyword matches one at a time instantly. Despite this advantage, and even though all trials were capped at ten minutes, the average trial in the base case scenario still took 189% as long as the average trial in the text analytics scenario.

The average relevancy score of the finished proposals in the base case scenario was 42%, whereas the average relevancy score of the finished proposals in the text analytics scenario was 75%. This difference was likely attributable to the ability of the virtual advisor to filter out irrelevant information. As the experiment was controlled to ensure the chemical engineers were facing issues outside their previous experiences, there was a possibility that participants could make an error in judgment thinking that certain information was relevant when it actually was not. In addition, some issues appear to be more challenging than do others, as evidenced by longer trial times and fewer proposals from proposers. The disproportionate number of outliers in the base case scenario suggests that more is going on than challenging questions. In the base case scenario, there were 21 total proposals made, but there were only 15 relevant proposals (71% of proposals made and 47% of total opportunities). In the text analytics scenario, there were 25 total proposals made, with 24 of them relevant (96% of proposals made and 86% of total opportunities). It is speculated that the natural language processing of the virtual advisor assisted in ensuring relevancy. One participant said, "With [the virtual advisor], when you can see the answers together, you can mentally group like with like. You know what to disregard, so you don't go down the wrong track."

In the base case scenario, the average accuracy score of relevant proposals was 50%. In the text analytics scenario, the average accuracy score of relevant proposals was 64%. Despite having four more total proposal opportunities than the text analytics scenario, the base case scenario yielded a much smaller sample size of relevant proposals, as 53% (17 out of 32) total proposal opportunities in the base case scenario did not convert to relevant proposals. A previous study involving the same virtual advisor found that subject matter experts considered the results of the advisor to be similar to the results returned by a human with experience in the domain (4). Like the higher percent relevance of proposals, the higher accuracy of relevant results in the text analytics scenario is also speculated to be the result of the natural language processing employed by this virtual advisor.

Conclusions

Notable in the results is the distinction between the augmentation of human performance and the replacement of humans. Intuitively, it makes sense that trained machines outperform humans in finding data across enormous and growing bodies of DUT files. Intuitive, it also makes sense that trained humans outperform machines at interpreting relevant information presented by trained machines. Text analytics appears to show strong potential as a valuable tool when applied to the chemical engineering domain. The virtual advisor in question emerges as apparently faster and more accurate than traditional information extraction methods. This appears to be a function of the effectiveness of the virtual advisor as well as a function of the ineffectiveness of incumbent solutions. This conclusion fits with other findings showing that human performance can be augmented in the industry by advanced data processing techniques (5). Given the improvements in efficiency and performance, the virtual advisor (and text analytics methods more generally) could generate enormous value in the chemical engineering domain in areas including increased safety, improved production, and reduced costs.

Additional testing is advisable. Most importantly, the focus group experiment could benefit from increased transparency on the questions asked and the participating individuals and companies. The focus group experiment could also benefit from a more tightly controlled corpus of data to ensure that accurate resolutions to the trial issues are contained within. Likewise, the experiment could benefit from a more standardized set of trial issues with measured attempts to control equality in their difficulty and commonality. A much larger set of proposers and trial issues and a more expert and assuredly impartial set of evaluators would further help to firm the conclusions made in this report. Furthermore, while the trials in the experiment were sourced from participants, and while each participant uniformly agreed that the trials resembled those commonplace within their organizations, more data confirming the specifics of why and how need to be collected from the organizations and any appropriate changes to the focus group setting duly incorporated. Finally, the study needs to be expanded to include other areas of the chemical engineering domain.

General research into the nature of what makes text analytics successful within a domain and comparative controlled evaluation of text analytics across multiple domains will complement focus group studies such as this. Larger scale studies likely require more concerted attention from across major professional societies linked to the industry.

References

1. Larsen, Åshild Hanne, Equinor CIO Keynote Address, *Society of Petroleum Engineers*, SPE Digital Transformation Study Group, Digital Transformation Annual Congress 2019, May 10, Houston, Texas, USA.
2. Feblowitz, Jill (2013), "Analytics in Oil and Gas: The Big Deal about Big Data," *Society of Petroleum Engineers*, SPE Digital Energy Conference, 5-7 March, The Woodlands, Texas, USA, SPE-163717-MS.
3. Jensen, Jesse W. and Perrons, Robert K. (2015), "Data as an asset: What the oil and gas industry can learn from other sectors about 'Big Data'," *Energy Policy*, Volume 81, June 2015, pp. 117-121.
4. Walker, Alec (2019), "Streamline digital information searches with advanced data techniques," *Hydrocarbon Processing*, January, pp. 23-24.
5. Sumbal, M., Tsui, E. and See-to, E. (2017), "Interrelationship between big data and knowledge management: an exploratory study in the oil and gas sector", Journal of Knowledge Management, Vol. 21 No. 1, pp. 180-196.

580204 Global Teams: Technology & People

Hear some of the practical aspects of working with teams spread across the world: time zones and cultures. This includes both cost effective technology solutions and the critical aspects of building and managing teams you do not see on a daily or even regular basis. Tips, tricks and key success factors will be presented and discussed.

581026 Digital Twin, Raising It out of the Realm of a Cliché

The Engineering and Industrial world are lagging manufacturing industries to truly embrace digitalization. AVEVA is accelerating the digital transformation journey of their customers and disrupting their current business processes to add value opportunities to their businesses. A common disruptive technology is the concept of Digital Twin. Through our work with our customers, we realize that believe to define a Digital Twin for the industrials we should draw a parallel to a the actual Human Twin. A twin has a shape, behavior, senses and needs to work for 100 years. The same is true for the digital twin of an industrial asset whether that is a an offshore platform, a refinery, a power generator or a ship. The Digital Twin of a plant is defined by engineering data (digital shape), engineering models (digital behavior), real-time series data (digital senses) and finally maintenance data (digital health). Once this data is connected just like a human it is can be stored in memory (historian), analyzed to make the right decisions (predictive analytics), learns from its mistakes (machine learning), asks questions (cognitive thinking) to make it more efficient (APC, real-time optimization), be mobile (smart devices) and finally visualize data in context (3D, X Reality etc...). AVEVA will present how it is connecting engineering, maintenance and operations data, people and business processes to bring the digital twin out of the realm of a Cliché.

580756 Asset Optimization Software Suites to Drive Digitalization – from Process Modeling to Optimization to Asset Optimization

Process Simulation has undergone significant transformation over the past decades driven by innovations in engineering best practices catalyzed by chemical engineering, mathematical optimization and software innovations. It has come a long way from automating the design calculations and process modeling to an environment for process optimization accelerating collaboration across the disciplines and tackling optimization across multiple dimensions simultaneously. The use of Process Simulation has also expanded from the traditional design applications to operations and planning. This digital transformation has created significant value in the industry resulting in capital and energy savings combined with better yields and capacities, increased safety and dramatic improvements in engineering productivity and production. Rapid advances in machine learning, advanced multivariate analytics, high performance mobile computing, and equipment and sensor connectivity are playing a role in enabling a whole new generation of technological solutions. The paper will cover the vision for Process Simulation including expansion from Process to Asset Optimization with first principles-based models and data-based machine learning models for process engineering, operations and performance management.

580855 A General Digital Applications Platform

Deep process knowledge captured within physics-based models is a key element for the successful digitalization of process industries. Digital applications making use of models derived from first principles are increasingly being used in industrial practice, particularly in the area of process operations. They include both open-loop applications, such as model-based soft sensing and monitoring of equipment degradation, and closed-loop ones such as Real-Time Optimization and nonlinear Model Predictive Control.

However, most non-trivial digital applications involve much more than a mathematical model being solved in a simulation or optimization mode. They often require multiple model-based calculations being scheduled over time and exchanging data with each other, while potentially being subject to occasional failures. They also involve extensive communication with external data servers, such as distributed control systems, plant historians, commercial databases and user dashboards. Moreover, the data involved in all such communications may be subject to systematic and/or random errors, and may even occasionally become unavailable.

This paper describes a recently developed general software platform for resilient and sustainable digital applications, taking account of the above considerations. The platform significantly reduces the cost and increases the reliability of development, testing, deployment and maintenance of diverse applications within a unified software architecture. Industrial examples illustrating the flexibility of the proposed design are presented.

580649 Concept of Digital Twin and Digital Thread in the Chemical & Process Industry

The vision of digital twins involves the digital twin of the product, the digital twin of process performance and the digital twin of operation.

In the R &D, stages product development and recipe transformation in with digital models can accelerate innovation. In design & engineering stages, all the disciplines involved in a plant project can generate a shared data model –saving valuable time and expenses and at the same time increasing engineering quality. The digital twin also allows the plant to be simulated prior to the critical commissioning phase, which ensures a reliable implementation. The operators can also receive targeted training before the system is commissioned. The digital twin of the operation significantly increases the reliability and availability of your plant by continuously monitoring your mechanical assets using IIoT.

The realization of this needs cross-organization collaboration. Realization of the concept needs various tools, software, hardware and collaboration platforms to work together. This talk with explore those possibilities and readiness.

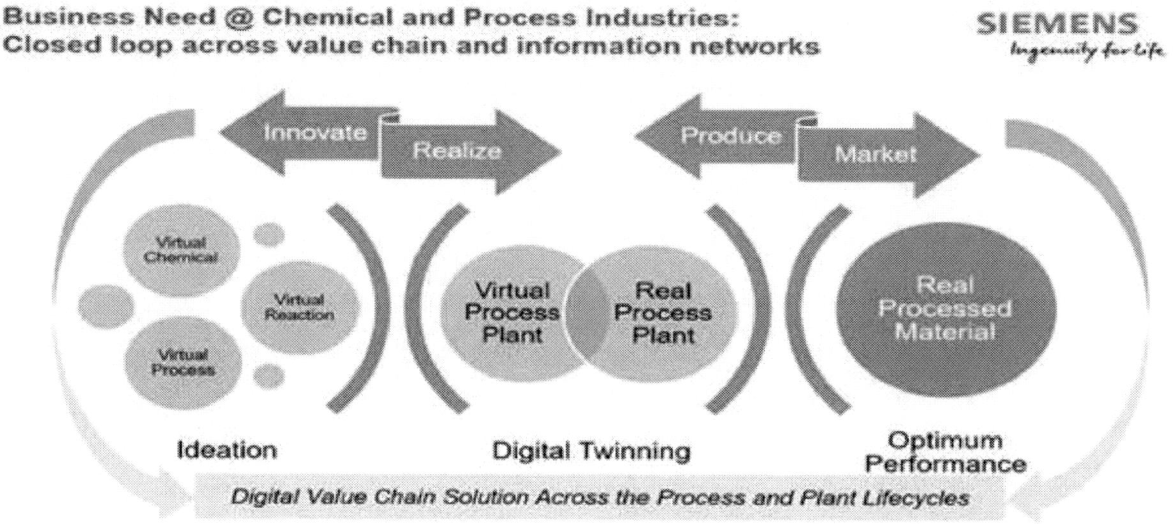

571319 Digital Twin for Aluminum Smelting

Building digital twins for manufacturing processes like aluminum smelting using first principles modeling has been considered challenging owing to the scale and complexity of such processes. A considerable mismatch is normally observed between measured data and predicted values of desired variables. This mismatch is usually because of uncertainty in the information fed to the model such as model assumptions, parameters, correlations from literature, etc. A systematic approach to identify such missing information would be to pose the model as an input-reconstruction problem. While a typical input reconstruction problem is solved using optimization-based algorithms where the objective would be to minimize the deviation between data and prediction, it may not be suitable for complex processes such as aluminum smelting. In this work we present a novel method to reconstruct missing information using a combination of first principles modeling and data analysis. The input reconstruction problem is reformulated as a set of differential algebraic equations (DAE) using measured data. Solving these DAEs help identify the missing input information and other unknown states in one shot. This method gives a complete data set (reconstructed inputs and measured data) which is used to train machine learning (ML) algorithms to build new correlations and predict unknown inputs directly in the digital twin model thereby giving more accurate predictions of desired process parameters.

580071 Ethical Dilemmas - an Introduction

This introduction will suggest different types of ethical considerations and will outline ethical dilemmas typically found in engineering education, government and business. In addition, results will be shared from the 2017 Ethics Survey of AIChE Members.

580126 Data: Yours? Mine? Ours? and When Is Enough Enough?

As engineers, we handle lots of data, interpret data and recommend actions based on data all with an ethical underpinning of safety for the public. We must consider regulations, business objectives, privacy and security concerns and more which can combine with human and organizational paradigms and objectives to create pressure or conflicts that result in ethical dilemmas. Our first case study will put participants in real life situations relevant to both the research environment and industry.

580230 Workshop--Part 1 "I Followed the Recipe, Where Is My Cake?": Important Questions to Ask in Considering Transition to Management

You are an individual contributor, and you aspire to become a leader. But have you thought about why you want to change to management and how one makes that transition? Most people deliberate over it for a long time and work towards this goal. The roles you take in the years before you want to make the jump will demonstrate that you are ready for management. This session will focus on steps to take and opportunities to target to prepare for the transition into management.

580231 Workshop--Part 2: I Know the Why. Now What?

If you know the why, the next thing on the list is to figure out what needs to be done to show that you are ready to be a manager. What should you be prepared for? Unfortunately, they don't teach us how to be a manager in school, and degrees don't necessarily land us a management job. In this workshop, we will list down some activities and opportunities that we can take to be able to demonstrate that you are prepared for management roles.

577933 Teamwork and Teamwork Evaluation Skills

Engineering curriculum is interspersed with opportunities for students to engage in teamwork. But often students are placed on teams with no instruction on how to be a successful member of a team. In addition, faculty aren't often trained on how to observe or evaluate teams. As a result, teamwork is often graded on the final deliverable of the team and not the teamwork skills that students should be developing.

In this talk the presenter will share favorite literature and resources on teamwork and teamwork evaluation skills. Examples of teamwork skills collected from the literature include establishing a team charter or team expectations, using team roles, and countering unproductive behaviors. Tips on observing and evaluating teams will also be shared.

562425 Using the Trolley Problem in Chemical Engineering Ethics Education

The so-called trolley problem has been used as a common tool of ethicists to introduce difficult topics of morality to the classroom and to the public. In this problem, and its many variants, participants are asked to participate in a choice that will either result in the hypothetical death of or injury to one person versus several people. In many instances, respondents have been found to gravitate to choosing harm to the individual over the group. However, depending on how the questions is posed and how the people involved are described, the majority of respondents may switch to favoring harm to the group. Using functional magnetic resonance imaging studies, researchers have found the decision between harm to the individual verses the group engage distinct regions of the human brain, and may reveal important moral decision-making processes.

These same processes are often in play in the daily life of a working chemical engineer, when attempting to assess the hazards and risks associated with their decisions in the design process or during daily operations. As automation increases, engineers are finding themselves in the position to code these ethical choices into the scripts needed to make decisions when a human operator cannot make them in real-time. As such, implications from these automated choices need to be considered well in advance by all engineers, from those designing the next autonomous vehicles, to those designing the next generation of oil refineries.

For ten years, we have used the trolley problem (and its variants) within the engineering ethics portions of our senior Unit Operations Laboratory to introduce concepts of "personal" verses "impersonal" ethical dilemmas. In this talk, we will discuss the results from the use of this common tool of ethicists, and how it can be used in a chemical engineering classroom settings to help students realize the influences on their ethical reasoning that often go unnoticed, even when they can dramatically alter the ultimate consequences to coworkers or the public.

578314 Preparing Students for Professional Success: Cross-Disciplinary Lessons

An important objective for many disciplines at academic institutions is to impart students with the specific skills and viewpoints needed to succeed in their chosen profession. Each discipline has carefully considered how their courses can best prepare students for professional life and has come up with strategies for ensuring their students practice the necessary skills. While in some cases the techniques used by a particular discipline are unique to that discipline, it is likely that there are commonalities between the approaches. It may be possible for educators in chemical engineering, for example, to learn from faculty in journalism on how to prepare students for the professional world. In this project, a survey was sent to faculty in 46 different departments at Kansas State University to evaluate how they taught students the skills they would need in their professional life. In addition, 33 different classes were visited, representing 23 different academic departments. This presentation will describe both quantitative and qualitative results on the ways in which faculty impart the skills their students need to succeed as professionals. Both commonalities and differences will be described, with specific attention to unique ways of training students that may be transferable between disciplines.

562135 A Novel Course Sequence on Critical Thinking, Publishing, and Publication Ethics for Engineering Graduate Students

We teach two, one-credit graduate courses for engineering (STEM) graduate students at the University of South Carolina aimed at improving the ability of graduate students to publish peer-reviewed papers and to do so while conforming to ethical standards of publication and research. The first course, which is for new graduate students, teaches critical reading and analysis of the research literature in their field. In this first course we also address the ethical issues of plagiarism and how to properly cite, paraphrase, and summarize information. The second course, for graduate students preparing to publish, teaches students how to write a journal paper or proposal for review. The ethical issues addressed in this second course are authorship, review standards and processes, and avoiding data falsification and data fabrication. The theoretical underpinning of both the reading and writing courses is the Paul-Elder model of critical thinking (Paul & Elder 2012).

We will present the critical thinking model along with the course materials including course syllabi, typical assignments, resources for publication ethics, and methods for effectively assessing student work. We will specifically discuss how the courses have been designed to work within the constraints of typical graduate degree program requirements and within the resource constraints of engineering graduate programs.

Paul R, Elder L. 2012. *Critical thinking : Tools for taking charge of your learning and your life*. Boston: Pearson. 493 pp.

AGILE MANAGEMENT OF PHD PROJECTS: A RISK-MITIGATION TOOL FOR GRADUATE EDUCATION

Laura Pirro, Ghent University, Ghent, Belgium
Marina Montero Carrero, Ristretto.black, Brussels, Belgium
Francesco Contino, Université Catholique de Louvain, Louvain-la-Neuve, Belgium
Joris W. Thybaut, Ghent University, Ghent, Belgium

Introduction

Pursuing a PhD can feel like walking on a tightrope: exhilaration and adrenaline arising from the challenge can drive one forward, but the fear of failure and of a ruinous fall might be paralyzing [1].

The psychological distress faced by graduate students has multiple roots, such as difficulties in maintaining a work-family balance, lack of inspirational guidance, uncertainties about future career outcomes, etc. [2]. Addressing, not to mention solving, these challenges cannot be done with a one-size-fits-all strategy; each PhD is a unique experience by a unique individual [3].

However, one common feature is shared by all research projects, namely the strive to push boundaries of science forward. This is a highly uncertain and sometimes unpredictable task and for this reason traditional management strategies, such as 'waterfall' methods [4] and Gantt charts [5], tend to fail. PhD projects do not follow the plans and so far there is no clear risk-mitigation system to support the graduate student when this happens.

An alternative, *Agile* philosophy for research management was already advocated a decade ago [6], but, being focused on the challenges and the structure of the software industry, it failed to hit the big time in Academia. More recently the implementation of an *Agile* methodological workflow is finally raising interest in Academic environments [7],[8]. However, the variety in activities and challenges to be tackled during a long-term research project, such as the ones carried out by PhD candidates, still requires a comprehensive assessment of the advantages and potential pitfalls of *Agile*.

In the present work an *Agile* approach focused on graduate education, in general, and the management of PhD projects, more specifically, is proposed. The goal is to draw the attention of a broad Academic audience to this management philosophy, as a promising instrument for potentially enhancing the productivity of research projects, while at the same time improving the psychosocial wellbeing of graduate candidates.

Agile workflow for research projects

When the *Agile* Manifesto [9] was first published, 18 years ago, it started a disruptive revolution in the world of software development. One of the core principles at the heart of this philosophy is 'Responding to change over following a plan'. Having an *Agile* mindset means being flexible, responsive, adaptable. In an *Agile* world, progress is measured through the attainment of concrete results and working outcomes, rather than by ticking off successive tasks on a to-do list. An early, limited result, which can be swiftly elaborated in a later stage, is more significant than a perfect result achieved only by the deadline of the project [10]: the first allows for feedback during the project, while the latter simply does not. It is tempting to transpose the characteristics of *Agile* management as described above to graduate research projects.

Based on the available literature [11], some guidelines could be proposed to translate the concepts of *Agile* into practical guidelines for the daily activities of PhD researchers [12]. The suggested steps to be implemented are the following:

1. A big block of work is divided into several 'layers' of activities. What defines a layer is that it ends with the attainment of a concrete result, as opposed to the traditional waterfall methodology where activities are organized purely based on the timeline. Each layer will be undertaken in a specific, limited period of time (e.g. 2-12 weeks), named a *sprint.*
2. *Sprint* planning: the PhD student, the supervisor and any other stakeholders (for example a postdoc, industrial partner, or masters student) meet for around 30 minutes aiming to set the goal of the sprint and its duration. Everybody must agree on these two points, to ensure that the entire research team is on the same page and that prospects are the same for everyone. As part of the sprint planning, the sprint review meeting (see step 5) can already be scheduled.
3. *Sprint* execution: work! Maximum focus is dedicated on each particular task for a definite amount of time.
4. Weekly '*scrum*': meeting between the PhD student and the supervisor for a maximum of 15 minutes, but ideally every week to keep the momentum, (e.g. the same timeslot every week, outside of 'conventional' working hours to make sure there are no meetings, teaching activities etc. which may get in the way). The point of this meeting is to be short and efficient: better avoiding chairs so that everyone stands up and there is no risk of getting too comfortable and overextending the meeting. No laptops, not even papers. Only three questions need to be answered:
    a. What was done the previous week in order to contribute to the goal?
    b. What will be done next week in order to contribute to the goal?
    c. Are there any impediments?
5. *Sprint review + sprint retrospective + sprint planning*: at the end of the sprint, a meeting is planned including the PhD student and all the stakeholders involved to discuss the obtained outcomes and whether those are in line with the goals and expectations set up in the previous planning (review). This is the moment to go into the specifics and brainstorm altogether. It is crucial to discuss the encountered difficulties, in order to make sure the next sprint is continuously improved compared to the previous one (retrospective). This is the phase when impediment removal (= problem solving, in *Agile* jargon) is dealt with: sincerity and transparency are key. As previously mentioned, *Agile* is all about adapting to change, and plans can change. Go back to point 1 and restart the planning, addressing the next layer of work in a new sprint.

The conceptual difference between a traditional and an *Agile* approach to PhD management is summarized in Figure 1.

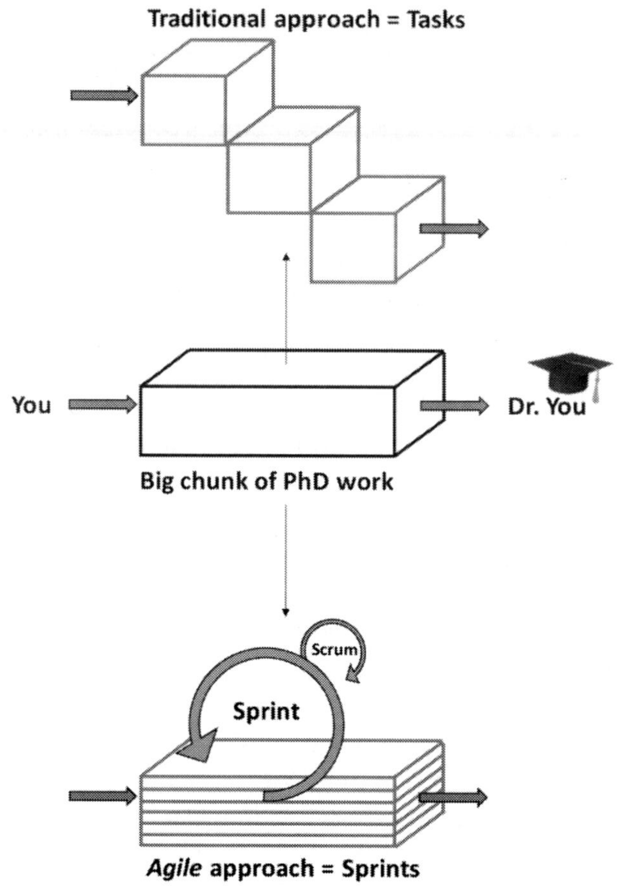

Figure 1. Conceptual difference between traditional PhD management (red) and *Agile* PhD management (green).

Academic *Agile* case studies

To illustrate the proposed approach, we provide a sample overview of the application of the *Agile* principles to three typical tasks of a researcher (for example, in chemical engineering). The simplified case studies reported in Table 1 offer a conceptual view of some activities which can be encountered in a research project and are meant to exemplify how the *Agile* approach aims at finalized, even if preliminary, results in the early stages of work.

Table 1. Comparison of traditional and *Agile* approach for three recurring case studies in graduate education of, among others, chemical engineers.

Long-term goal	Traditional approach			*Agile* approach		
	Scope of task 1	Scope of task 2	Scope of task 3	Goal of sprint 1	Goal of sprint 2	Goal of sprint 3
Writing a manuscript	Write 'Methodology' and 'Results'	Write 'Discussion' and 'Conclusions'	Write 'Introduction'	Write down the key concepts of each section as bullet points; sketch figures of the 'expected' results	Add actual figures and tables with captions; include main references in each section	Elaborate text around the bullet points and the figures
Building a simulation code	Write detailed subroutines for input/output	Write detailed subroutines for calculations	Test the code	Write and test simplified input/output/ calculation subroutines	Increase the level of detail of the main subroutines and get them to work independently	Integrate all the subroutines in one working code; testing
Performing an experimental campaign	Design of Experiments (DoE) over the full range of variables and operating conditions	Execution of all the experiments	Data processing	Simplified DoE, execution of selected experiments and data processing	Increase the number of factors in the DoE, execution of new experiments, data merging and processing	Increase the number of levels per factor in the DoE, execution of new experiments, data merging and processing

In order to move towards realistic, applied case studies, we recruited researchers, students and professors willing to apply the proposed *Agile* workflow to a short-term project [13]. By remotely helping the participants to set-up the case studies and following up the progress vis Skype updates, several collaborations with other Academic institutions were established. Three examples are described below.

Writing a research proposal - University of North Carolina at Chapel Hill (May – August '19)

Who: Bailey DeBarmore is a senior PhD Student at the department of Epidemiology of the Gillings School of Global Public Health, under the supervision of Prof. Wayne Rosamond, who is also responsible for educational activities in the same department.

What: Bailey had approximately 14 weeks to draft and submit a full research proposal to obtain funding for her dissertation project. She had to start from the literature review, then move into developing and refining the specific aims, all the way down to a finalized text.

Why *Agile:* Driven by the desire of reaching her goal in the most efficient way, Bailey recognized that the *Agile* principles could respond to her need of a structured working method, while at the same time retaining a certain flexibility, as required by innovative research projects.

How: Bailey split her work in 4 sprints, which were guided by Wayne in weekly scrum meetings of approx. 30 min. For each of her sprints, she set a finalized written document as goal to achieve, continuously building up on previously written text and transitioning, as of the first moment, from a draft to a finalized proposal.

Feedback: -) No negative points were highlighted by Bailey and Wayne.
+) The case study was considered successful, with the proposal being handed in timely before the deadline. Wayne could appreciate a considerable difference compared to previous students he had coached in the past. In fact, as opposed to 'just' brainstorming for long time, Bailey's *Agile* approach gave him the possibility to always discuss and provide feedback over a concrete document.
Both Bailey and Wayne confirmed that they are likely to implement *Agile* in their future work.

Summer project with bachelors students – University of California, San Diego (June – August '19)

Who: Bill Connacher is a PhD student in materials science and engineering at the Jacobs School of Engineering – UCSD, in prof. James Friend's lab. Cesar Collins and Aneth Martinez worked under Bill's supervision at UCSD as part of the ENLACE 2019 summer research program for high school and college students.

What: Cesar and Aneth had to design and manufacture a capillary wicking structure within a Surface Acoustic Wave (SAW) device and had approx. 7 weeks to accomplish their goal.

Why *Agile*: Bill had acted as mentor for undergraduate students in the past and had encountered some difficulties in facilitating their achievement of a finalized result. In order to avoid the pitfall of uncompleted outcomes at the end of the internship, Bill decided to adopt an *Agile* workflow in mentoring Cesar and Aneth.

How: 3 sprints, which were managed via daily scrum meetings. The complexity of the work in each of the sprints was gradually increased, starting from a simplified laser ablation system, which, due to originally unforeseen project requirements, was later on upgraded with an additional capability.

Feedback: -) Bill encountered considerable difficulties in keeping the scrum meetings short and only once per day. This was attributed to the total lack of previous knowledge of the students about the research topic and it highlights one of the major requirements/drawbacks of *Agile* implementation: the need of a team which is highly skilled in the field. It can be expected that, in some cases, this might hinder adoption of *Agile* by starting PhD students and or early-stage researchers.
+) As opposed to previous internships, Cesar and Aneth's work resulted in a finalized structure which, despite not being optimized, was capable of effectively transporting fluid. Bill appreciated the opportunities arising from *Agile*, such as the incremental approach in defining goals and the flexibility in adjusting expectations, and he will try to implement a comparable workflow in some aspects on his own research.

Computational Fluid Dynamics simulations – CORIA, University of Rouen (June – September '19)

Who: Julien Reveillon is professor of Computational Fluid Dynamics (CFD) at the University of Rouen Normandy and researcher at UMR 6614-CORIA in the field of multiphase flows. Cléante Langrée is a PhD student, funded by INERIS, working under Julien's supervision. Javier Anez is a postdoctoral researcher affiliated to the CNRS, collaborating in Julien's group. Nathan Mauger is a graduate student at ESITech Rouen and he carried out a summer internship in CORIA with guidance from Javier and Julien.

What: Cléante and Julien: comparison of premixed flames CFD solvers, using flame velocity, flame thickness and adiabatic temperature as outputs of interest from 3D Large Eddy Simulations;
Javier and Nathan: numerical simulations of a boat's hull movement with various water aeration rates, with a specific focus on boat's drag and stability issues.

Why *Agile*: Julien's commitment to innovation in teaching and learning sparked his interest towards *Agile*. His concern is that too little time and efforts are dedicated to the development of professional skills and working ethics in engineering graduate programs. He saw *Agile* as an opportunity to challenge himself and his co-workers with an alternative, and potentially promising, approach to research.

How: In both cases, 5 sprints with a duration of 2 weeks each, were foreseen and managed via 1-2 scrum meetings per week. Each of the sprints had three goals: solve at least one of the several numerical challenges, write/update a written report on the achieved progress, summarize the sprint results into a structured presentation.

Feedback: -) Estimation of the time which would be actually required for a given task was often arduous. As a consequence, this could lead to demotivation and struggle. This is one of the most problematic features of research work, since the issues which are being tackled in research projects are usually new and/or not fully understood, and it represents an onerous challenge for any management style. Nathan found that he could partially mitigate this difficulty with the *Agile* workflow, by re-negotiating the goal of successive sprints in the review and planning phase.
+) In both cases an improvement in the mentoring relationships was observed, both from a professional and a personal point of view. Frequent interactions, even if shorter, resulted in enhanced communication and overall higher efficiency in the work. Enforcing the writing/presentation since from the first moment, each sprint was always concluded with a finalized work which served as solid starting point for the next one. Julien and Javier will continue their journey into Agile with their future projects. Having experienced the benefits in his own work, Nathan, who will supervise an undergraduate project during his upcoming final year as grad student, will try and adopt an *Agile* coaching style.

Conclusions

The application of an *Agile* management style to research projects could positively impact the Academic environment in terms of increased flexibility and adaptation to the ever-changing challenges to be tackled in science. Particularly, graduate students and their supervisors are encouraged to follow this approach as risk-mitigation plan to face the endeavors of PhD projects.

The case studies carried out so far were evaluated only qualitatively. However, the participants reported an increase in productivity, a higher success rate in the achievement of project deliverables, and, thanks to the enhanced number of contact moments, an improved perception of the working environment. More importantly, all the involved participants confirmed to be willing to further adopt *Agile* in the upcoming research challenges that they will be confronted with.

The promising results from finalized and ongoing case studies in different research domains encourage the establishment of additional collaborations with other Academic institutions. These collaborations will aim at quantitatively evaluating an *Agile* management of research projects on a statistically relevant number of subjects.

Acknowledgments

The authors would like to thank L. Comegna and D. Otyuskaya for the useful discussions in the initial stage of this work, and the members of the CaRE group at the Laboratory for Chemical Technology (University of Ghent) for being the first 'guinea pigs' of the proposed approach.

The authors are particularly grateful to all the participants of the currently ongoing seven case studies for their trust, time and dedication to the project.

References

1. C. Woolston, Graduate Survey: a Love-Hurt Relationship, Nature (2017)
2. K. Levecque, F. Anseel, A. De Beuckelaer, J. Van der Heyden, L. Gisle, Work Organization and Mental Health Problems in PhD Students, Research Policy (2017)
3. L. Constantin, How to handle the one-size-fits-all PhD, Nature Career Column (2018)
4. C.D. Tupper, Data Organization Practices, Data Architecture: from Zen to Reality, Morgan Kaufmann (2011)
5. H.L. Gantt, Work, wages and profit, Engineering Magazine (1910)
6. X. Amatriain Rubio, https://www.slideshare.net/xamat/agile-science (2008)
7. L. May and T. Runyon, LabScrum: A Case Study For Agility in Academic Research Labs, PsyArXiv (2019)
8. O. Bergman, https://crastina.se/agile-project-management-taught-to-students-at-karolinska-institutet/ (2019)
9. Manifesto for Agile Software Development (2001)
10. J. Sutherland, SCRUM The Art of Doing Twice the Work in Half the Time, Random House Business Books (2015)
11. Agile Project Management: Quick Start Guide, Albany (2016)
12. L. Pirro, How agile project management can work for your research, Nature Career Column (2019)
13. M. Montero Carrero, Ready to try Agile for research? Join our upcoming study, Ristretto.black (2019)

569949 TA Survival Guide: Navigating Your First Week of Classes

The first day walking into the class room as a teaching assistant can be scary, but with the right preparation, you can be successful in the classroom. There are a few steps that you can take to make your first week of classes as painless as possible. This can include talking to your instructor to negotiate your role in the classroom, understanding the tools available to you, and learning how to gain confidence and rapport with your students. In this workshop, you will learn how to navigate all of these obstacles and how to successfully ensure that you will have a productive and successful semester as a TA. You will be able to meet your developmental goals, be an asset to your instructor, and be a proponent for student success.

580292 Background and Discussion about Retaining Women Engineers

Dr. Urmila Diwekar is currently President of the Stochastic Research Technologies LLC, and the Vishwamitra Research Institute (VRI, www.vri-custom.org), a non-profit research organization that she founded in 2004 to pursue multidisciplinary research in the areas of Optimization under Uncertainty and Computer-aided Design applied to Energy, Environment, and Sustainability.

From 2002-2004, she was a Professor in the Departments of Chemical Engineering, Bio-Engineering, and Industrial Engineering, and in the Institute for Environmental Science and Policy, at the University of Illinois at Chicago (UIC). From 1991-2002 she was on the faculty of the Carnegie Mellon University, with early promotions to both the Associate and the Full Professor level.

In chemical engineering, she has worked extensively in the areas of simulation, design, optimization, control, stochastic modeling, and synthesis of chemical processes. She has made major contributions to research on batch distillation, and this work is well recognized. She is the author of more than 165 peer-reviewed research papers, 7 books, and 15 chapters, and has given over 350 presentations and seminars, and has chaired numerous sessions in national and international meetings. In separations, she is well-known for her pioneering work in batch distillation design, simulation, optimization, and optimal control.

She has published more than 31 peer-reviewed journal papers, 65 presentations (15 plenary/keynote presentations), and conducted 11 educational workshops on this topic. She has also published the first book on this topic. This is an authoritative, first of its kind book which is widely used all over the world. This book received excellent reviews and the second edition of this book was published in early 2012. She is the author of a comprehensive software package for batch distillation simulation, optimal design, and control called MultiBatchDS which is used in pharma and specialty chemical companies in USA, Europe, and Asia.

In this session, she will give her personal opinion about how to maintain women in the industry.

580293 Background and Discussion about Retaining Chemical Engineering Women in Industry

Gayle J. Gibson led the DuPont Engineering Research and Technology organization of internal engineering consultants and the company's Field Engineering Program for early-career engineers.

Gayle joined DuPont in 1983 and held a variety of roles through which she implemented best practices and enabled global business revitalization on behalf of multiple, high-performing business units. Her contributions included engineering design, process engineering, research and development, product and process development, manufacturing supervision, market development, business strategy, growth and supply chain improvement.

From 2008-2013, Gayle was Director – Corporate Operations for the Chair of the Board of Directors and Chief Executive Officer of DuPont. In that role she facilitated processes and engagement among key DuPont operating and leadership teams. This included working across businesses, functions and regions to drive alignment and clarity on priorities, advancing core values and resolving issues in line with the company's mission and strategic objectives. She also supported the Chief Executive Officer as co-lead of the National Academy of Engineering project "Changing the Conversation – from Research to Action" to improve the public understanding of engineering.

A certified Six Sigma Champion, Gayle holds a Bachelor of Science in chemical engineering from Texas A&M University and a Master of Business Administration from the European University. She is a member of AIChE, Tau Beta Pi and the Society of Women Engineers

In this session, she will be giving her personal thoughts on how to retain women engineers in the industry.

580291 Background and Discussion about Retaining Women in Industry

Ana Davis heads Health, Safety, and Environment (HSE) in North America for Syngenta, a global agriculture company. She manages all of Syngenta's corporate HSE programs, which include active ingredient manufacturing, product formulation, seeds operations, and commercial sales. Her team develops tools to improve safety culture across the company.

Ana began her career as a process control engineer with Exxon, later working as a facilities engineer for Advanced Micro Devices. Prior to joining Syngenta in 2009, she held a variety of production and HSE management roles with Kao Specialties Americas. Before that, she was a plant manager at Dow and held several positions with Union Carbide Corp., becoming a plant production manager after the Dow Chemical merger.

A native of Spain, Ana was recognized in 2005 with the Hispanic Engineer National Achievement Award for Professional Achievement in Industry. She serves on the advisory boards of AIChE's Center for Chemical Process Safety (CCPS) and North Carolina State Univ.'s Dept. of Chemical and Biomolecular Engineering. Ana received her BS degree in chemical engineering from North Carolina State Univ., where she and her husband have established an endowment for chemical engineering undergraduate students.

In this session Ana will give her personal point of view about maintaining women engineers in the industry.

587477 Background and Discussion about Retaining Women Engineers in Industry

Bill serves as Director – Engineering Technology for the Chemours Company. He is accountable for leading the company's efforts to transform and renew the leveraged engineering competencies in support of Manufacturing Operations, large Capital Growth Projects, Sustainability and Process Safety.

Bill has over 30 years of experience with DuPont and Chemours in R&D, Engineering and Manufacturing technology management, as a well having assignments in corporate risk management and trade secret protection.

Bill graduated with a B.S. degree in Chemical Engineering from Columbia University and received his M.S.C.E.P. and Ph.D. degrees in Chemical Engineering from the Massachusetts Institute of Technology.

580356 My Management Journey, from Technology Developer to Innovation Leader

An innovation leader embodies domain expertise, perseverance, risk-tolerance, creativity, adaptability and trustworthiness, with a strong scientific and engineering foundation. Innovation leaders believe the right science applied in the right way can impact people worldwide, generating breakthroughs that make life better. At 3M, my teams leverage 51 technology platforms to solve customers' problems. I have experienced first-hand the joy of discovery, the power of curiosity and the importance of determination to deliver new technologies and products to diverse markets and improve lives. I will describe experiences and moments of pride for my team's achievements.

581071 Impacts on Management from an Evolving World and Profession

Chemical engineering and the world are evolving; how does that affect management? In some essential ways, not at all: A decision is still a decision, a smile is still a smile as time goes by. Communicating the organization's mission clearly is still essential. So is rewarding people who perform well, not the ones who don't.

In other ways, everything has changed: Staff are expected to be reachable instantly, travel for meetings is slashed in favor of virtual "face-to-face" and audio meetings, and detailed management data on budgets, supply chains, and production are available instantaneously for analytics and decision-making - at least in principle. None of that is chemical-engineering-specific.

The profession-driven changes for chemical engineers are more subtle. The pace of mergers and company re-organizations is due in part to diverse opportunities in material, bioscience, and information technologies. Similarly, employees themselves have more diverse expectations. Perhaps more than in any other engineering discipline, education of chemical engineers is to learn a core of expertise, not a specific job. In recent years, students also learn more about the diverse ways they'll apply their education, and they often look forward to being entrepreneurial. That presents opportunities and challenges for their managers.

581072 Advancing Technology Though Measurement Science

The mission of the National Institute of Standards and Technology (NIST) is to promote U.S. innovation and industrial competitiveness by advancing measurement science, standards, and technology in ways that enhance economic security and improve our quality of life. We do this by working in close partnership with US industry to build the technical infrastructure needed for advancing technology. In this presentation, I will highlight successful examples pf this partnership from the nSoft consortium and the Materials Genome Initiative.

581073 Engineering and Science at the National Laboratories

The National Laboratories represent complex organizations that conduct forefront research, and that house some of the most advanced technological instruments available to the scientific community. In this presentation, I will present a brief overview of Argonne National Laboratory and of Fermilab, and how industry and academia interact with these institutions to pursue large multidisciplinary projects involving large teams of scientists and engineers. Importantly, the success of these projects depends on robust management structures that have evolved over the years to enable and facilitate scientific research. I will describe some of these structures, along with some of their advantages and limitations.

580300 Part 1 – Applying Creativity Tools

In **Part One** of the workshop (approximately 75 minutes), you will learn how to jumpstart your creativity with a simple exercise that yields a high number of ideas for further exploration.

580301 Part 2 – Practicing Master Minding

In **Part Two** of the workshop (45 minutes), we will test drive a sample advisory board by brainstorming solutions to the challenges introduced in Part One.

580302 Part 3 – Setting & Achieving Goals

Finally, we will conclude with **Part Three** (30 minutes) of the workshop in which you'll learn how to design achievable goals with accountability from your leadership advisory board.

With a mix of learning and hands-on practice, you will learn how to apply five leadership mindsets to help you achieve higher levels of collaboration within your teams while building a toolkit for fast-tracking your career success. Moreover, these techniques will help you balance work, leisure, health, and relationships so that you will feel more joy and purpose as a multi-dimensional chemical engineering leader. If you are interested in accelerating your career as a leader in your organization, this workshop is for you!

578258 Practicing the ABCs of Technical Communication Using Realistic Problems and Homework Memos

Practicing solving problems through homework is an important aspect of building students' technical skill set. However, the format of these problems and the way they are reported do not generally align with the way problems are framed or reported on outside of the classroom. I have recently instituted a new type of homework problem where students are assigned realistic problems and instead of only turning in the handwritten or spreadsheet calculations, they must also submit a short memo reporting on their work in the context of a fictional company. While the memo format is not necessary, these style assignments give students practice with the ABCs of technical communication: Accuracy, Brevity, and Clarity. Through these memos, students learn to identify the key concepts and results from their work, communicate information concisely, and manipulate and present data in a manner that is accurate and intelligible. I will demonstrate example assignments from two classes: an elective on alternative energy and a core class on chemical engineering and the environment. In the alternative energy class, where some students had minimal technical writing experience, there was an observable improvement in written communication skills over the six memo assignments given that semester.

574688 Flipping Writing Revisions: Students Devise Self-Revision through Reflection in the Engineering Laboratory

A balance is yet to be struck between writing to learn and communication in the workplace across the engineering curriculum. Most engineering students find writing reports in core courses, including laboratory sequences, challenging or a task of reporting 'just the data', presenting a conundrum to instructors: How do we support engineering students to think deeply about core concepts and discipline-specific discourse simultaneously? Engineering faculty who teach laboratory often find grading to be burdensome, time-consuming, and unsustainable in the long-term. Furthermore, industry members want engineers who can communicate across different audiences in the corporate domain. This study focuses on a small group reflective writing activity designed with several goals in mind: sustainability and promoting critical self-assessment through reflection, a tool well-acknowledged as key to student learning and professional identity development. Participants include students (N=52) enrolled in a junior-level Chemical Engineering Laboratory II course at a large minority-serving research university in the American Southwest. A core English writing instructor was embedded in the course to assist students with written communication to a broader audience across the boundaries of the discipline. There were three major student deliverables in the laboratory: two individual short reports and one oral presentation per team. The reflection activity with each team was facilitated in the second round of short reports. We conducted mixed methods analysis on the student work. Individual student reflections were assessed by all the course instructors using the following criteria: indication of deep thought which translated into an audience-minded revision plan or sufficient argument against revision and specific evidence cited to support their perspective on revision. The activity was designed as a guided reflection on each section of the report reminding students that a report is a narrative, and the content of each section should advance that narrative. Higher scores were recorded for student reflections that included specific evidence from the report itself and direct language about readers or audience needs. Reflections that neglected to mention the reader's needs, or the writer's impact on the reader, and read like a to-do list received lower scores than reflections that mentioned one or more of the criteria. We found the reflections tended to fall into three categories: defense of rhetorical choices, fix errors like a to-do list, and audience-minded revision plans. Students' ability to have adaptable writing styles for different audiences was also evidenced by their citing of real world applications as broader context for an experiment in the report. The reflection activity provides students the opportunity to go beyond a typical checklist of report corrections to self-revising with audience in mind.

575419 Utilizing Peer Review to Improve the Quality of Unit Operations Assignments

After graduation, we quickly learn that technical communication requires a significant time commitment and many iterations on the same draft. Unfortunately, undergraduate students rarely have the time to go through the iterative process of perfecting their communication skills. This limits not only the quality of their assignments, but also their ability to learn successful strategies for improving communication. While allowing students the opportunity to revise communication assignments based on faculty feedback would be ideal, the huge time commitment associated with grading and providing feedback is limiting. Therefore, this presentation will focus on strategies for integration of peer review processes for both written and oral communication assignments into the unit operations lab. In going through the peer review process, students are required to submit initial drafts as early as a week prior to the deadline for the assignment. This gives them significant time to improve the assignment on their own, as well as integrate feedback from at least one peer. In submission of their final draft, students are asked to include a letter to the editor which highlights major revisions that were made based on both peer feedback, as well as faculty feedback from prior assignments. This holds students accountable for both reading and applying faculty feedback as they continue throughout a course. Since introducing the peer review process in the lab, we have seen significant improvement in the quality of communication assignments, especially with regard to minor mistakes that can be time-consuming to address when grading. As a result, grading can be completed more quickly with a focus on higher-level feedback.

564655 Communication Skills Instruction in a Two-Semester Capstone Design Course Sequence

The authors serve as instructors for the Chemical Process Design II (CHEN 4316) and Design III (CHEN 4317) course sequence in chemical engineering at Texas A&M University-Kingsville (TAMUK). An important aspect of these two courses is instruction and mentoring of students in the development of their communication skills. The two principal reasons that communication skills are taught in this course sequence are (a) a necessity to prepare students for a successful engineering career, and (b) to satisfy the ABET student outcome 'g' (new ABET student outcome '3'), which is "an ability to communicate effectively with a range of audiences". Additionally, to meet a SACS (Southern Association of Colleges and Schools) requirement, the Design III course is designated as "writing intensive" for the chemical engineering undergraduate degree.

Prior to this design course sequence, undergraduate chemical engineering majors have completed three courses related to communication skills, specifically two courses in undergraduate English Rhetoric and Composition, and a Communications course in Professional Communication. Upon entering the capstone design sequence, most students indicate they do not yet feel comfortable with their communication skills, in particular their oral communication skills. Therefore, both oral and written communication skills are emphasized throughout both of the courses in this sequence. One group oral presentation is required in Design II, while four group oral presentations are required in Design III. For all of these presentations, individual feedback is provided to the students on their presentation skills. Additionally, the instructors will also go over presentation do's and don'ts as a series of talks occurs over several class days. Three written group project reports are required in Design II. In the Design III course, the writing requirements are greater because of the writing intensive aspect of the course—ten individual writing assignments, along with a group mid-term report and a group final design report. At least two of these written submittals in Design III require draft and final versions, with the students incorporating changes according to instructor comments for the final. This presentation focuses on the successes and challenges associated with the intensive communication instruction in this capstone design course sequence at TAMUK.

566319 Framework for Conducting a Technical Communication Crosswalk across a Chemical Engineering Curriculum

ABET student outcome 4 states that undergraduate students must have "an ability to communicate effectively with a range of audiences" upon graduation. To meet this student outcome, and enhance the communication skills of students, an undergraduate engineering program can intentionally place technical communication instruction across its curriculum. However, not all engineering programs intentionally scaffold experiences to develop technical communication skills through the curriculum. To assess the number, placement, and quality of communication-related graded events within our chemical engineering curriculum, we developed a technical communications crosswalk framework. The framework included mapping of written communication events (research papers, contemporary issue papers, capstone design reports, resume writing exercises, and short technical communications), oral presentations, and laboratory reports from freshman through senior year in a total of 16 required courses for our chemical engineering majors. Linkages for each type of technical communication experience to experiences in course pre-requisites were also included to determine opportunities for connecting events across courses. A comprehensive, multiple-part survey was used to collect characteristics of each technical communication event, to include, the topic, whether the event was completed by individual students or in a team, the percentage of the total course grade the event was worth, and where the graded event was placed within the course, amongst others. Instructor assessment of each event was also collected for courses taught by chemical engineering faculty (e.g., mass and energy balances, chemical reaction engineering, etc.). The technical communications crosswalk framework provided insight into the way our program addresses ABET student outcome 4 and identified some areas for possible improvement. For example, we are considering how to improve technical communications early in the student experience, which could include developing a 1.0 credit technical communications seminar taken in the sophomore year. While the results will vary, the crosswalk framework introduced in this study can be applied to engineering programs at many universities.

564578 Theory, Design, and Early Implementation Results of a Multi-Tiered, Scaffolded Writing Pedagogy

As undergraduate engineering students graduate and advance in their careers, they are faced with multiple tasks that require them to write extensively, whether that be in industry or graduate school. Because of the insufficient writing requirements in undergraduate engineering curricula, many engineers are unprepared for the writing-demand necessary to convey their own ideas or understanding of ideas. More rigorous writing practices would not only improve effective communication skills as undergraduate students pursue their education, but this can also help students develop a deeper conceptual foundation of engineering topics. The writing pedagogy of interest follows a scaffolded approach and was implemented alongside a current hands-on learning pedagogy in a 46-student, junior-level fluid mechanics class (CHE 332) at Washington State University. Students began by taking a pre-test with a descriptive paragraph of the technical phenomena occurring in a Venturi meter; certain phrases of the paragraph were designated as having two options, where they had to choose the correct phrase. Students then used the Venturi hands-on learning device, followed by a posttest. In the posttest, the sections that were previously presented in the form of two options were changed to blanks, where students were required to write their own comprehensible phrases. The results of this implementation were compared to a previous semester when the scaffolded writing pedagogy was not used, with that cohort of students acting as the control group. From pre- to posttest, it was found that the experimental group performed significantly better than the control group with 99% confidence and a medium effect size, based on ranges for partial eta squared. These initial steps are intended to help students learn how to discuss engineering concepts, leading them to properly write full paragraphs on their own on the final exam in early-May. The results of the final exam will be included in the presentation.

579139 Findings from MTO Commercialization – Don't Forget about the Small Stuff

In the rush to commercialize a new technology, there are many things to consider: reactor scale up & design; engineering design of major equipment; simulations; thermodynamics; sales tools; customer discussions, and a lot more. One of the things that can be easily missed is what happens to components that are produced in minor concentrations. It may be easy to ignore these when a simulation indicates that they will be washed away in waste streams (or maybe become product impurities). However, these minor concentration components can become problematic when they don't respond exactly as predicted, especially when they build up their concentrations in recycle loops. This talk will discuss some these type of issues that have occurred with MTO commercialization and what we are doing about them.

573116 De-Risking Technology Development and Scale-up Efforts Effectively in Fuel and Chemical Processes

The sequential pathway for de-risking new technologies includes bench scale testing, pilot programs, demonstration plants and finally commercial plant construction. In the past, hastening, or eliminating any of the above steps would increase the risk to novel technologies, and to the companies that invest in them. Unfortunately, this pathway requires significant time and financial investment. In the past few years, in response to increased pressure to speed up the pathway, engineers have developed many tools and strategies for mitigating that risk. Modern process simulation, targeted piloting programs, toll processors, directional technology development plans and modular strategies can all help companies achieve commercial success from technology development through scale-up to commercialization. When employed properly, these, and other tools can decrease project risk, shorten the time to commercialization, and limit the overall cost of developing innovative technologies. This presentation will discuss how to utilize these, and other tools to quickly and effectively de-risk process technology development. Using case studies of past projects, it will also review the process of developing technologies from bench scale through pilot and demonstration scale to full commercialization within the framework of these de-risking tools. Case studies include next-generation biodiesel development, conversion of natural gas liquids into valuable chemicals and marine fuel, and production of high purity alumina from waste streams.

565316 Can Outsourcing R&D Balance Myopic Business Management?

Abstract

When Ford Motor Co.'s CEO, Jim Hacket, announced plans in April 2018 to reduce spending by $25.5 billion over the next four years [1], Ford staff were concerned about how the pending cuts would impact R&D investments [2]. Similarly, DowDuPont Inc.'s CEO, Ed Breen, announced the elimination of R&D spending on "moonshots" to focus on projects that cost less than $30 million to complete and, perhaps most importantly, carry a lower risk profile [3]. From one perspective, these announcements are merely part of normal operations, as business leaders make decisions to increase, reduce, or cut spending every day. On the other hand, these announcements raise a few questions:

- Were these part of a trend in R&D spending?
- What are some of the factors that affect CEO decision-making?
- What are some consequences of myopic decision-making?
- Can outsourcing R&D balance myopic business management?

After briefly addressing these questions, this presentation will disclose a relevant case study in which MATRIC, a contract R&D organization, successfully helped a client to develop and commercialize a new technology.

References

[1] A.Levine-Weinberg, "Ford Reveals More Details on its Massive Cost-Cutting Plan," Fox Business, August 13, 2018 (https://www.foxbusiness.com/markets/ford-reveals-more-details-on-its-massive-cost-cutting-plan).

[2] L.Brooke, "Underway on Nuclear Power," Automotive Engineering, June 2018, p. 18 (http://www.nxtbook.com/nxtbooks/sae/18AUTP06/index.php#/22).

[3] J.Kaskey, "DowDuPont CEO Kills Off R&D 'Moonshots' in Bid for Higher Profit," Bloomberg, May 31, 2018 (https://www.bloomberg.com/news/articles/2018-05-31/dowdupont-s-new-profit-formula-killing-more-research-projects).

555687 Succeeding at Innovation: Generating Value from New Technology at Each Stage of Development and Implementation

Succeeding at Innovation: Generating Value from New Technology at Each Stage of Development and Innovation

Technology provides increased capability, energizes our imagination, and leads to business advantages. Too often our embrace of technology is weakened by narrow and inaccurate measures of value, the challenge of changing existing practices, and the inertia of our own minds and customs.

Examining the effect new technology has on our systems at each stage of its development and implementation reveals the full value of its effect. The potential gains at each stage, how to map an uncertain path to the future, and methods to decide the level and timing of investment are discussed in the context of drug product delivery, processing, and scale up. Examples from modeling and simulation, 3-D printing, and novel lyophilization technology are used as illustrations.

552804 Heat Transfer Coefficient in Two Phase Liquid-Solid Fluidized Bed

Heat Transfer Coefficient in Two Phase Liquid-Solid Fluidized bed

Marwa Sabah Mahdi ,Dr.Jamal M.Ali* and Dr.Ameer A.Abdul-Rahman

University of Technology Department of chemical Engineering, Iraq

***80002@uotechnology.edu.iq**

Abstract

Hydrodynamic of liquid,-solid fluidized bed was studied experimentally as well as theoretically.The impact of the superficial velocities of liquid, particle diameter ,the liquid holdup, solid holdup, height of bed and minimum velocity fluidizing were investigated. Individual phase holdups were determined through static pressure approach from the information of the pressure drop, the bed height and the properties of the solid and liquid.

Heat transfer coefficients to a liquid-solid fluidized bed in a cylindrical tube have been measured using water as liquid phase and three sizes of spherical glass beads particles as solid phase and electrically heated single vertical tube heated electrically with different heat transfer rate (500, 600 and 700W). The independent variables included heat flux, liquid velocity and particle physical properties. The results showed that the heat transfer coefficient increased with increasing fluidizing water velocity and also with increasing heat flux. It was also determined that the heat transfer coefficient increased with increasing the diameter of solid particles and decreased with increasing the voidage fraction. On the other-hand, the heat transfer coefficient changed with the change of positions of heater distance from the distributer. The experimental results as well as a data bank containing a large number of measured heat transfer coefficients for solid- liquid fluidization over a wide range of operational parameter have been compared with the predictions of most published correlations. The results were agreed in compare with the existing correlations predictions. These correlations were found to be approximate in reproducing the data of present work.

Key word: liquid-solid, fluidized bed, minimum fluidization velocity- solid hold up, heat transfer

AUTHOR INDEX